Busy Student's

Anatomy & Physiology

Study Journal

AF797817

Reece Biomedical Consulting LLC
Fayetteville, NY

Margaret T. Reece/Reece Biomedical Consulting LLC
4977 Highbridge Street
Fayetteville, NY 13066
MedicalScienceNavigator.com

Book Layout ©2019 by D'vorah Lansky
VibrantMarketingPublications.com

Cover Photo by Mopic/Shutterstock.com

ISBN 978-0-9963513-3-1

Great things are done by a series of small things brought together.

—VINCENT VAN GOGH

How to Use This Journal

In your hands, you have a 30-lecture journal to help you organize your study of topics presented in your anatomy and physiology course.

The first step toward getting an *A* in A&P is to become clear about details of your class syllabus. This journal's first section helps you do that.

The second section has four blank calendars to cover the months in which your course is taught. Fill in the blanks with the proper months, year and dates. Add your exam dates from your syllabus.

The third section contains your journaling pages. There are four journal pages for each lecture. On the day of the lecture begin with page one. List major topics covered and any questions you have.

On the second and third journal pages for that lecture fill in dates to return there for review. Add those dates to your calendar to help you remember them. The fourth journal page for each lecture is blank for creating sketches to link ideas.

Check your calendar every day, and return to the journal pages for the lectures listed there.

As you review your previous lectures, jot down your new thoughts and realizations about the topics discussed.

When psychologists studied how the human mind learns best, they discovered that spaced review of new ideas enhanced learning. Use this journal as a tool to organize your study of A&P in a way that takes advantage of this new awareness of how mental processing works.

Frequently Asked Questions

Question: What if my exam for a lecture is scheduled before I finish its 30 day review?

Answer: Maintain your review timetable no matter when your exams are scheduled.

Anatomy and physiology is a cumulative science taught as individual systems. However, all body systems dynamically interact with each other at all times. The interleaving of spaced review of individual lectures has the added advantage of displaying how different body systems relate to each other. For example, muscle names refer to names of their attached bones. Nerves, arteries and veins are named for the muscles they supply.

Question: Why should I decide only 3-4 major topics for each lecture?

Answer: Each lecture presents a lot of detail. Your effort to sort through it and reduce the facts to 3 major themes forces your mind to connect pieces of a puzzle. This is a skill that will serve you well on exams that include short answer questions.

Question: Why is there a page for sketches?

Answer: Drawings help organize thoughts.

Question: How much extra time will journaling take?

Answer: Journaling itself should only take a few minutes a day. Thinking about what to write will become an ongoing process in your mind.

My Syllabus

For Anatomy &Physiology

In this section you will become fully acquainted with the information in your class syllabus about:

- Class Rules
- Grading
- Exam Format
- Missing Lecture/Lab
- And More

My Syllabus For A&P

Instructor:____________________ **Date:**__________

Why understanding my course syllabus is important to me:

____

____

____

____

Syllabus to do list:

- ❑ Find a quiet place to read my syllabus
- ❑ Take note when my instructor mentions it
- ❑ Review dates of exams and quizzes
- ❑ Notice drop-without-grade deadline
- ❑ Be clear about points for *A, B,* or *C* grade
- ❑ Find out what happens if I miss an exam
- ❑ Learn rules about missing lectures/lab
- ❑ Make note of penalty for being late to lab
- ❑ Understand rules about cell phone use
- ❑ Study lab safety rules
- ❑ Keep syllabus handy and read often

Questions I Still Have

About class rules:

About missing lecture, quizzes or lab:

About grading:

About exam format:

My Calendar

For Anatomy & Physiology

This section has four blank monthly calendars. Fill in the year, months and dates for the period of your anatomy and physiology course.

Within the box for each date, there is a place to insert lecture numbers for review. You will add these as you begin the four journal pages for each lecture.

The left side of the calendar has a column for you to add the dates of your exams.

Dates For Reviewing Lectures

	Sunday	Monday	Tuesday
Month: ___	___ Lecture#___ Lecture#___ Lecture#___ Lecture#___	___ Lecture#___ Lecture#___ Lecture#___ Lecture#___	___ Lecture#___ Lecture#___ Lecture#___ Lecture#___
Exams: ___ ___ ___ ___	___ Lecture#___ Lecture#___ Lecture#___ Lecture#___	___ Lecture#___ Lecture#___ Lecture#___ Lecture#___	___ Lecture#___ Lecture#___ Lecture#___ Lecture#___
___ ___ ___ ___ ___	___ Lecture#___ Lecture#___ Lecture#___ Lecture#___	___ Lecture#___ Lecture#___ Lecture#___ Lecture#___	___ Lecture#___ Lecture#___ Lecture#___ Lecture#___
___ ___ ___ ___ ___	___ Lecture#___ Lecture#___ Lecture#___ Lecture#___	___ Lecture#___ Lecture#___ Lecture#___ Lecture#___	___ Lecture#___ Lecture#___ Lecture#___ Lecture#___
___ ___ ___ ___ ___	___ Lecture#___ Lecture#___ Lecture#___ Lecture#___	___ Lecture#___ Lecture#___ Lecture#___ Lecture#___	___ Lecture#___ Lecture#___ Lecture#___ Lecture#___

Wednesday	**Thursday**	**Friday**	**Saturday**
______ Lecture#__ Lecture#__ Lecture#__ Lecture#__	______ Lecture#__ Lecture#__ Lecture#__ Lecture#__	______ Lecture#__ Lecture#__ Lecture#__ Lecture#__	______ Lecture#___ Lecture#___ Lecture#___ Lecture#___
______ Lecture#___ Lecture#___ Lecture#___ Lecture#___	______ Lecture#___ Lecture#___ Lecture#___ Lecture#___	______ Lecture#___ Lecture#___ Lecture#___ Lecture#___	______ Lecture#___ Lecture#___ Lecture#___ Lecture#___
______ Lecture#___ Lecture#___ Lecture#___ Lecture#___	______ Lecture#___ Lecture#___ Lecture#___ Lecture#___	______ Lecture#___ Lecture#___ Lecture#___ Lecture#___	______ Lecture#___ Lecture#___ Lecture#___ Lecture#___
______ Lecture#___ Lecture#___ Lecture#___ Lecture#___	______ Lecture#___ Lecture#___ Lecture#___ Lecture#___	______ Lecture#___ Lecture#___ Lecture#___ Lecture#___	______ Lecture#___ Lecture#___ Lecture#___ Lecture#___
______ Lecture#___ Lecture#___ Lecture#___ Lecture#___	______ Lecture#___ Lecture#___ Lecture#___ Lecture#___	______ Lecture#___ Lecture#___ Lecture#___ Lecture#___	______ Lecture#___ Lecture#___ Lecture#___ Lecture#___

Dates For Reviewing Lectures

	Sunday	Monday	Tuesday
Month: ________	______ Lecture#___ Lecture#___ Lecture#___ Lecture#___	______ Lecture#___ Lecture#___ Lecture#___ Lecture#___	______ Lecture#___ Lecture#___ Lecture#___ Lecture#___
Exams: ________ ________ ________ ________	______ Lecture#___ Lecture#___ Lecture#___ Lecture#___	______ Lecture#___ Lecture#___ Lecture#___ Lecture#___	______ Lecture#___ Lecture#___ Lecture#___ Lecture#___
________ ________ ________ ________ ________	______ Lecture#___ Lecture#___ Lecture#___ Lecture#___	______ Lecture#___ Lecture#___ Lecture#___ Lecture#___	______ Lecture#___ Lecture#___ Lecture#___ Lecture#___
________ ________ ________ ________ ________	______ Lecture#___ Lecture#___ Lecture#___ Lecture#___	______ Lecture#___ Lecture#___ Lecture#___ Lecture#___	______ Lecture#___ Lecture#___ Lecture#___ Lecture#___
________ ________ ________ ________ ________	______ Lecture#___ Lecture#___ Lecture#___ Lecture#___	______ Lecture#___ Lecture#___ Lecture#___ Lecture#___	______ Lecture#___ Lecture#___ Lecture#___ Lecture#___

Wednesday	**Thursday**	**Friday**	**Saturday**
______ Lecture#__ Lecture#__ Lecture#__ Lecture#__	______ Lecture#__ Lecture#__ Lecture#__ Lecture#__	______ Lecture#__ Lecture#__ Lecture#__ Lecture#__	______ Lecture#___ Lecture#___ Lecture#___ Lecture#___
______ Lecture#___ Lecture#___ Lecture#___ Lecture#___	______ Lecture#___ Lecture#___ Lecture#___ Lecture#___	______ Lecture#___ Lecture#___ Lecture#___ Lecture#___	______ Lecture#___ Lecture#___ Lecture#___ Lecture#___
______ Lecture#___ Lecture#___ Lecture#___ Lecture#___	______ Lecture#___ Lecture#___ Lecture#___ Lecture#___	______ Lecture#___ Lecture#___ Lecture#___ Lecture#___	______ Lecture#___ Lecture#___ Lecture#___ Lecture#___
______ Lecture#___ Lecture#___ Lecture#___ Lecture#___	______ Lecture#___ Lecture#___ Lecture#___ Lecture#___	______ Lecture#___ Lecture#___ Lecture#___ Lecture#___	______ Lecture#___ Lecture#___ Lecture#___ Lecture#___
______ Lecture#___ Lecture#___ Lecture#___ Lecture#___	______ Lecture#___ Lecture#___ Lecture#___ Lecture#___	______ Lecture#___ Lecture#___ Lecture#___ Lecture#___	______ Lecture#___ Lecture#___ Lecture#___ Lecture#___

Dates For Reviewing Lectures

	Sunday	Monday	Tuesday
Month: ________	______ Lecture#___ Lecture#___ Lecture#___ Lecture#___	______ Lecture#___ Lecture#___ Lecture#___ Lecture#___	______ Lecture#___ Lecture#___ Lecture#___ Lecture#___
Exams: ________ ________ ________ ________	______ Lecture#___ Lecture#___ Lecture#___ Lecture#___	______ Lecture#___ Lecture#___ Lecture#___ Lecture#___	______ Lecture#___ Lecture#___ Lecture#___ Lecture#___
________ ________ ________ ________ ________	______ Lecture#___ Lecture#___ Lecture#___ Lecture#___	______ Lecture#___ Lecture#___ Lecture#___ Lecture#___	______ Lecture#___ Lecture#___ Lecture#___ Lecture#___
________ ________ ________ ________ ________	______ Lecture#___ Lecture#___ Lecture#___ Lecture#___	______ Lecture#___ Lecture#___ Lecture#___ Lecture#___	______ Lecture#___ Lecture#___ Lecture#___ Lecture#___
________ ________ ________ ________ ________	______ Lecture#___ Lecture#___ Lecture#___ Lecture#___	______ Lecture#___ Lecture#___ Lecture#___ Lecture#___	______ Lecture#___ Lecture#___ Lecture#___ Lecture#___

Wednesday	Thursday	Friday	Saturday
______ Lecture#__ Lecture#__ Lecture#__ Lecture#__	______ Lecture#__ Lecture#__ Lecture#__ Lecture#__	______ Lecture#__ Lecture#__ Lecture#__ Lecture#__	______ Lecture#___ Lecture#___ Lecture#___ Lecture#___
______ Lecture#___ Lecture#___ Lecture#___ Lecture#___	______ Lecture#___ Lecture#___ Lecture#___ Lecture#___	______ Lecture#___ Lecture#___ Lecture#___ Lecture#___	______ Lecture#___ Lecture#___ Lecture#___ Lecture#___
______ Lecture#___ Lecture#___ Lecture#___ Lecture#___	______ Lecture#___ Lecture#___ Lecture#___ Lecture#___	______ Lecture#___ Lecture#___ Lecture#___ Lecture#___	______ Lecture#___ Lecture#___ Lecture#___ Lecture#___
______ Lecture#___ Lecture#___ Lecture#___ Lecture#___	______ Lecture#___ Lecture#___ Lecture#___ Lecture#___	______ Lecture#___ Lecture#___ Lecture#___ Lecture#___	______ Lecture#___ Lecture#___ Lecture#___ Lecture#___
______ Lecture#___ Lecture#___ Lecture#___ Lecture#___	______ Lecture#___ Lecture#___ Lecture#___ Lecture#___	______ Lecture#___ Lecture#___ Lecture#___ Lecture#___	______ Lecture#___ Lecture#___ Lecture#___ Lecture#___

Dates For Reviewing Lectures

	Sunday	Monday	Tuesday
Month: ________	______ Lecture#___ Lecture#___ Lecture#___ Lecture#___	______ Lecture#___ Lecture#___ Lecture#___ Lecture#___	______ Lecture#___ Lecture#___ Lecture#___ Lecture#___
Exams: ________ ________ ________ ________	______ Lecture#___ Lecture#___ Lecture#___ Lecture#___	______ Lecture#___ Lecture#___ Lecture#___ Lecture#___	______ Lecture#___ Lecture#___ Lecture#___ Lecture#___
________ ________ ________ ________ ________	______ Lecture#___ Lecture#___ Lecture#___ Lecture#___	______ Lecture#___ Lecture#___ Lecture#___ Lecture#___	______ Lecture#___ Lecture#___ Lecture#___ Lecture#___
________ ________ ________ ________ ________	______ Lecture#___ Lecture#___ Lecture#___ Lecture#___	______ Lecture#___ Lecture#___ Lecture#___ Lecture#___	______ Lecture#___ Lecture#___ Lecture#___ Lecture#___
________ ________ ________ ________ ________	______ Lecture#___ Lecture#___ Lecture#___ Lecture#___	______ Lecture#___ Lecture#___ Lecture#___ Lecture#___	______ Lecture#___ Lecture#___ Lecture#___ Lecture#___

Wednesday	Thursday	Friday	Saturday
______ Lecture#__ Lecture#__ Lecture#__ Lecture#__	______ Lecture#__ Lecture#__ Lecture#__ Lecture#__	______ Lecture#__ Lecture#__ Lecture#__ Lecture#__	______ Lecture#___ Lecture#___ Lecture#___ Lecture#___
______ Lecture#___ Lecture#___ Lecture#___ Lecture#___	______ Lecture#___ Lecture#___ Lecture#___ Lecture#___	______ Lecture#___ Lecture#___ Lecture#___ Lecture#___	______ Lecture#___ Lecture#___ Lecture#___ Lecture#___
______ Lecture#___ Lecture#___ Lecture#___ Lecture#___	______ Lecture#___ Lecture#___ Lecture#___ Lecture#___	______ Lecture#___ Lecture#___ Lecture#___ Lecture#___	______ Lecture#___ Lecture#___ Lecture#___ Lecture#___
______ Lecture#___ Lecture#___ Lecture#___ Lecture#___	______ Lecture#___ Lecture#___ Lecture#___ Lecture#___	______ Lecture#___ Lecture#___ Lecture#___ Lecture#___	______ Lecture#___ Lecture#___ Lecture#___ Lecture#___
______ Lecture#___ Lecture#___ Lecture#___ Lecture#___	______ Lecture#___ Lecture#___ Lecture#___ Lecture#___	______ Lecture#___ Lecture#___ Lecture#___ Lecture#___	______ Lecture#___ Lecture#___ Lecture#___ Lecture#___

My Journal

Semester______ Course#_______

To keep your study plan on track, use 5-10 minutes after each lecture to jot down 3-4 major topics of the lecture. What was this lecture about?

Then, compose 2-3 questions about the topics presented. Be brief.

If you do not have 10 minutes after class, do this as soon as you can on the day of the lecture.

Add review dates to your next two journaling pages for that lecture and add those dates to your calendar.

Add any sketches to the fourth journaling page that you think will jog your memory when you return to review this lecture.

Check your calendar every day to see which lectures are scheduled for review.

When reviewing a lecture, list your new ideas and realizations about the topics covered.

A&P Lecture # ____ Date: ____

3-4 Major Topics Covered

My Questions About These Topics Are:

Review A&P Lecture #______

Ideas and Realizations

Next Day, Date: ______________________________

Ideas and Realizations

8 Days After Lecture, Date: ____________________

Review A&P Lecture #______

Ideas and Realizations

15 Days After Lecture, Date: ______________________

____

____

____

____

____

____

Ideas and Realizations

30 Days After Lecture, Date: ______________________

____

____

____

____

____

____

Sketches

A&P Lecture # ____ Date: ____

3-4 Major Topics Covered

My Questions About These Topics Are:

Review A&P Lecture #______

Ideas and Realizations

Next Day, Date: ________________________

Ideas and Realizations

8 Days After Lecture, Date: __________________

Review A&P Lecture #______

Ideas and Realizations

15 Days After Lecture, Date: ____________________

Ideas and Realizations

30 Days After Lecture, Date: ____________________

Sketches

A&P Lecture # ____ Date: ____

3-4 Major Topics Covered

My Questions About These Topics Are:

Review A&P Lecture #______

Ideas and Realizations

Next Day, Date: ______________________________

____

____

____

____

____

____

Ideas and Realizations

8 Days After Lecture, Date: ____________________

____

____

____

____

____

____

Review A&P Lecture #______

Ideas and Realizations

15 Days After Lecture, Date: ____________________

Ideas and Realizations

30 Days After Lecture, Date: ____________________

Sketches

A&P Lecture # ____ Date: ____

3-4 Major Topics Covered

My Questions About These Topics Are:

Review A&P Lecture #______

Ideas and Realizations

Next Day, Date: ______________________________

____

____

____

____

____

____

Ideas and Realizations

8 Days After Lecture, Date: ____________________

____

____

____

____

____

____

Review A&P Lecture #______

Ideas and Realizations

15 Days After Lecture, Date: ____________________

Ideas and Realizations

30 Days After Lecture, Date: ____________________

Sketches

A&P Lecture # ____ Date: ____

3-4 Major Topics Covered

My Questions About These Topics Are:

Review A&P Lecture #______

Ideas and Realizations

Next Day, Date: ______________________________

Ideas and Realizations

8 Days After Lecture, Date: ____________________

Review A&P Lecture #______

Ideas and Realizations

15 Days After Lecture, Date: ____________________

____

____

____

____

____

____

Ideas and Realizations

30 Days After Lecture, Date: ____________________

____

____

____

____

____

____

Sketches

A&P Lecture # ____ Date: ____

3-4 Major Topics Covered

My Questions About These Topics Are:

Review A&P Lecture #______

Ideas and Realizations

Next Day, Date: ______________________________

Ideas and Realizations

8 Days After Lecture, Date: ______________________

Review A&P Lecture #______

Ideas and Realizations

15 Days After Lecture, Date: ____________________

____

____

____

____

____

____

Ideas and Realizations

30 Days After Lecture, Date: ____________________

____

____

____

____

____

____

Sketches

A&P Lecture # ____ Date: ____

3-4 Major Topics Covered

My Questions About These Topics Are:

Review A&P Lecture #______

Ideas and Realizations

Next Day, Date: ______________________________

____

____

____

____

____

____

Ideas and Realizations

8 Days After Lecture, Date: ____________________

____

____

____

____

____

____

Review A&P Lecture #______

Ideas and Realizations

15 Days After Lecture, Date: ______________________

Ideas and Realizations

30 Days After Lecture, Date: ______________________

Sketches

A&P Lecture # ____ Date: ____

3-4 Major Topics Covered

My Questions About These Topics Are:

Review A&P Lecture #______

Ideas and Realizations

Next Day, Date: ______________________________

____

____

____

____

____

____

Ideas and Realizations

8 Days After Lecture, Date: ____________________

____

____

____

____

____

____

Review A&P Lecture #______

Ideas and Realizations

15 Days After Lecture, Date: ____________________

____

____

____

____

____

____

Ideas and Realizations

30 Days After Lecture, Date: ____________________

____

____

____

____

____

____

Sketches

A&P Lecture # ____ Date: ____

3-4 Major Topics Covered

My Questions About These Topics Are:

Review A&P Lecture #______

Ideas and Realizations

Next Day, Date: ____________________________

____

____

____

____

____

____

Ideas and Realizations

8 Days After Lecture, Date: ____________________

____

____

____

____

____

____

Review A&P Lecture #______

Ideas and Realizations

15 Days After Lecture, Date: ____________________

Ideas and Realizations

30 Days After Lecture, Date: ____________________

Sketches

A&P Lecture # ____ Date: ____

3-4 Major Topics Covered

My Questions About These Topics Are:

Review A&P Lecture #______

Ideas and Realizations

Next Day, Date: ______________________________

____

____

____

____

____

____

Ideas and Realizations

8 Days After Lecture, Date: ____________________

____

____

____

____

____

____

Review A&P Lecture #______

Ideas and Realizations

15 Days After Lecture, Date: ____________________

Ideas and Realizations

30 Days After Lecture, Date: ____________________

Sketches

A&P Lecture # ____ Date: ____

3-4 Major Topics Covered

My Questions About These Topics Are:

Review A&P Lecture #______

Ideas and Realizations

Next Day, Date: ____________________________

Ideas and Realizations

8 Days After Lecture, Date: ____________________

Review A&P Lecture #______

Ideas and Realizations

15 Days After Lecture, Date: ____________________

____

____

____

____

____

____

Ideas and Realizations

30 Days After Lecture, Date: ____________________

____

____

____

____

____

____

Sketches

A&P Lecture # ____ Date: ____

3-4 Major Topics Covered

My Questions About These Topics Are:

Review A&P Lecture #______

Ideas and Realizations

Next Day, Date: ______________________________

____

____

____

____

____

____

Ideas and Realizations

8 Days After Lecture, Date: ____________________

____

____

____

____

____

____

Review A&P Lecture #______

Ideas and Realizations

15 Days After Lecture, Date: ____________________

Ideas and Realizations

30 Days After Lecture, Date: ____________________

Sketches

A&P Lecture # ____ Date: ____

3-4 Major Topics Covered

My Questions About These Topics Are:

Review A&P Lecture #______

Ideas and Realizations

Next Day, Date: ______________________________

Ideas and Realizations

8 Days After Lecture, Date: ____________________

Review A&P Lecture #______

Ideas and Realizations

15 Days After Lecture, Date: ____________________

Ideas and Realizations

30 Days After Lecture, Date: ____________________

Sketches

A&P Lecture # ____ Date: ____

3-4 Major Topics Covered

My Questions About These Topics Are:

Review A&P Lecture #______

Ideas and Realizations

Next Day, Date: ______________________________

Ideas and Realizations

8 Days After Lecture, Date: ____________________

Review A&P Lecture #______

Ideas and Realizations

15 Days After Lecture, Date: ____________________

Ideas and Realizations

30 Days After Lecture, Date: ____________________

Sketches

A&P Lecture # ____ Date: ____

3-4 Major Topics Covered

My Questions About These Topics Are:

Review A&P Lecture #______

Ideas and Realizations

Next Day, Date: ____________________________

Ideas and Realizations

8 Days After Lecture, Date: ____________________

Review A&P Lecture #______

Ideas and Realizations

15 Days After Lecture, Date: ____________________

____

____

____

____

____

____

Ideas and Realizations

30 Days After Lecture, Date: ____________________

____

____

____

____

____

____

Sketches

A&P Lecture # ____ Date: ____

3-4 Major Topics Covered

My Questions About These Topics Are:

Review A&P Lecture #______

Ideas and Realizations

Next Day, Date: ______________________________

____

____

____

____

____

____

Ideas and Realizations

8 Days After Lecture, Date: ____________________

____

____

____

____

____

____

Review A&P Lecture #______

Ideas and Realizations

15 Days After Lecture, Date: ____________________

____

____

____

____

____

____

Ideas and Realizations

30 Days After Lecture, Date: ____________________

____

____

____

____

____

____

Sketches

A&P Lecture # ____ Date: ____

3-4 Major Topics Covered

My Questions About These Topics Are:

Review A&P Lecture #______

Ideas and Realizations

Next Day, Date: ____________________________

Ideas and Realizations

8 Days After Lecture, Date: ____________________

Review A&P Lecture #______

Ideas and Realizations

15 Days After Lecture, Date: ____________________

Ideas and Realizations

30 Days After Lecture, Date: ____________________

Sketches

A&P Lecture # ____ Date: ____

3-4 Major Topics Covered

My Questions About These Topics Are:

Review A&P Lecture #______

Ideas and Realizations

Next Day, Date: ______________________________

Ideas and Realizations

8 Days After Lecture, Date: ____________________

Review A&P Lecture #______

Ideas and Realizations

15 Days After Lecture, Date: ____________________

____

____

____

____

____

____

Ideas and Realizations

30 Days After Lecture, Date: ____________________

____

____

____

____

____

____

Sketches

A&P Lecture # ____ Date: ____

3-4 Major Topics Covered

My Questions About These Topics Are:

Review A&P Lecture #______

Ideas and Realizations

Next Day, Date: ________________________

Ideas and Realizations

8 Days After Lecture, Date: __________________

Review A&P Lecture #______

Ideas and Realizations

15 Days After Lecture, Date: ____________________

Ideas and Realizations

30 Days After Lecture, Date: ____________________

Sketches

A&P Lecture # ____ Date: ____

3-4 Major Topics Covered

My Questions About These Topics Are:

Review A&P Lecture #______

Ideas and Realizations

Next Day, Date: ______________________________

____

____

____

____

____

____

Ideas and Realizations

8 Days After Lecture, Date: ____________________

____

____

____

____

____

____

Review A&P Lecture #______

Ideas and Realizations

15 Days After Lecture, Date: ____________________

____

____

____

____

____

____

Ideas and Realizations

30 Days After Lecture, Date: ____________________

____

____

____

____

____

____

Sketches

A&P Lecture # ____ Date: ____

3-4 Major Topics Covered

My Questions About These Topics Are:

Review A&P Lecture #______

Ideas and Realizations

Next Day, Date: ______________________________

____

____

____

____

____

____

Ideas and Realizations

8 Days After Lecture, Date: ____________________

____

____

____

____

____

____

Review A&P Lecture #______

Ideas and Realizations

15 Days After Lecture, Date: ____________________

Ideas and Realizations

30 Days After Lecture, Date: ____________________

Sketches

A&P Lecture # ____ Date: ____

3-4 Major Topics Covered

My Questions About These Topics Are:

Review A&P Lecture #______

Ideas and Realizations

Next Day, Date: ______________________________

Ideas and Realizations

8 Days After Lecture, Date: ____________________

Review A&P Lecture #______

Ideas and Realizations

15 Days After Lecture, Date: ____________________

____

____

____

____

____

____

Ideas and Realizations

30 Days After Lecture, Date: ____________________

____

____

____

____

____

____

Sketches

A&P Lecture # ____ Date: ____

3-4 Major Topics Covered

My Questions About These Topics Are:

Review A&P Lecture #_____

Ideas and Realizations

Next Day, Date: ____________________________

Ideas and Realizations

8 Days After Lecture, Date: ____________________

Review A&P Lecture #______

Ideas and Realizations

15 Days After Lecture, Date: ____________________

Ideas and Realizations

30 Days After Lecture, Date: ____________________

Sketches

A&P Lecture # ____ Date: ____

3-4 Major Topics Covered

My Questions About These Topics Are:

Review A&P Lecture #______

Ideas and Realizations

Next Day, Date: ______________________________

Ideas and Realizations

8 Days After Lecture, Date: ____________________

Review A&P Lecture #______

Ideas and Realizations

15 Days After Lecture, Date: ____________________

Ideas and Realizations

30 Days After Lecture, Date: ____________________

Sketches

A&P Lecture # ____ Date: ____

3-4 Major Topics Covered

My Questions About These Topics Are:

Review A&P Lecture #______

Ideas and Realizations

Next Day, Date: ______________________________

Ideas and Realizations

8 Days After Lecture, Date: ______________________

Review A&P Lecture #______

Ideas and Realizations

15 Days After Lecture, Date: ____________________

Ideas and Realizations

30 Days After Lecture, Date: ____________________

Sketches

A&P Lecture # ____ Date: ____

3-4 Major Topics Covered

My Questions About These Topics Are:

Review A&P Lecture #______

Ideas and Realizations

Next Day, Date: ______________________________

Ideas and Realizations

8 Days After Lecture, Date: ____________________

Review A&P Lecture #______

Ideas and Realizations

15 Days After Lecture, Date: ____________________

Ideas and Realizations

30 Days After Lecture, Date: ____________________

Sketches

A&P Lecture # ____ Date: ____

3-4 Major Topics Covered

My Questions About These Topics Are:

Review A&P Lecture #______

Ideas and Realizations

Next Day, Date: ________________________________

____

____

____

____

____

____

Ideas and Realizations

8 Days After Lecture, Date: ______________________

____

____

____

____

____

____

Review A&P Lecture #______

Ideas and Realizations

15 Days After Lecture, Date: ____________________

Ideas and Realizations

30 Days After Lecture, Date: ____________________

Sketches

A&P Lecture # ____ Date: ____

3-4 Major Topics Covered

My Questions About These Topics Are:

Review A&P Lecture #______

Ideas and Realizations

Next Day, Date: ________________________________

____

____

____

____

____

____

Ideas and Realizations

8 Days After Lecture, Date: ______________________

____

____

____

____

____

____

Review A&P Lecture #______

Ideas and Realizations

15 Days After Lecture, Date: ____________________

____

____

____

____

____

____

Ideas and Realizations

30 Days After Lecture, Date: ____________________

____

____

____

____

____

____

Sketches

A&P Lecture # ____ Date: ____

3-4 Major Topics Covered

My Questions About These Topics Are:

Review A&P Lecture #______

Ideas and Realizations

Next Day, Date: ______________________________

Ideas and Realizations

8 Days After Lecture, Date: ____________________

Review A&P Lecture #______

Ideas and Realizations

15 Days After Lecture, Date: ____________________

Ideas and Realizations

30 Days After Lecture, Date: ____________________

Sketches

A&P Lecture # ____ Date: ____

3-4 Major Topics Covered

My Questions About These Topics Are:

Review A&P Lecture #______

Ideas and Realizations

Next Day, Date: ________________________________

Ideas and Realizations

8 Days After Lecture, Date: ______________________

Review A&P Lecture #______

Ideas and Realizations

15 Days After Lecture, Date: _______________________

____

____

____

____

____

____

Ideas and Realizations

30 Days After Lecture, Date: _______________________

____

____

____

____

____

____

Sketches

Wrapping Up

Your Study of A&P

Take a few moments to appraise your progress.

Think about how journaling improved (or did not improve) your ability to learn anatomy and physiology.

Reflection

How did spacing my review of course lectures in this way help me learn anatomy and physiology?

____

____

____

____

What do I know now about studying anatomy and physiology that I didn't know when I began the course?

____

____

____

____

Did I really get better than expected test scores and grades using this study plan? Why do I think that?

____

____

____

____

Major Realizations

About The Author

Margaret T. Reece Ph.D. of Reece Biomedical Consulting LLC has been teaching anatomy and physiology to college students for over 30 years.

With her website MedicalScienceNavigator.com, books and online course Dr. Reece helps students achieve the *A* grade required for acceptance into the professional school they desire.

She has taught at 2-year and 4- year colleges, and served as a mentor to undergraduate, graduate and doctoral students.

Her books are available and may be previewed at:

amazon.com/author/margaretreece

Made in the USA
Monee, IL
04 November 2026

67377188R00085